YOUR KNOWLEDGE HAS VALUE

- We will publish your bachelor's and
  master's thesis, essays and papers

- Your own eBook and book -
  sold worldwide in all relevant shops

- Earn money with each sale

Upload your text at www.GRIN.com
and publish for free

Amalia Aventurin

Subgrain Size Piezometric on a Halite

GRIN Verlag

Bibliografische Information der Deutschen Nationalbibliothek:

Die Deutsche Bibliothek verzeichnet diese Publikation in der Deutschen National-
bibliografie; detaillierte bibliografische Daten sind im Internet über http://dnb.d-
nb.de/ abrufbar.

Imprint:

Copyright © 2013 GRIN Verlag GmbH
Druck und Bindung: Books on Demand GmbH, Norderstedt Germany
ISBN: 978-3-656-64470-5

This book at GRIN:

http://www.grin.com/en/e-book/272601/subgrain-size-piezometric-on-a-halite

Subgrain Size Piezometric on a Halite

I. Description of the used method

The Subgrain Size Piezometric is done by a Triassic halite from the upper bunter formation in Hengelo, Netherlands. The rock salt halite is white and normally in thin sections are no crystal boundaries observable, because of the cubic crystal structure and its optical isotropy. To make even the sub grain boundaries visible, the sample has to prepared in a special way, where the crystals appear blue and the boundaries white.

For the Subgrain Size Piezometric we uploaded first a picture 430.9-1, which we took as an example, in the program "ImageJ" and changed the type to 8-bit. The next step is adjusting the threshold. The program allows an automatically threshold, but for a better analysis it does be done manually. Figure 1 shows the converted image. Other programs for example MATLAB or ArcGIS can be used as well.

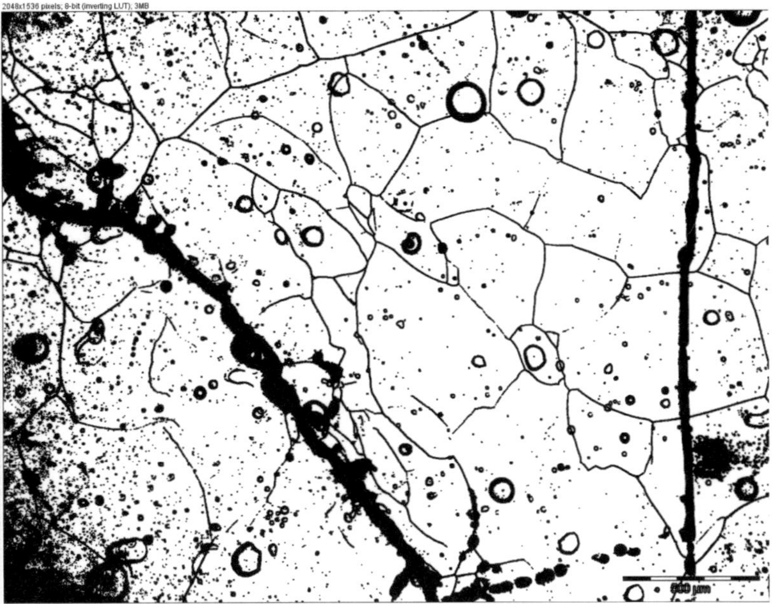

Fig. 1: Converted threshold of picture 430.9-1.

The aim of adjusting the threshold is to create the sub grain boundaries most visible. For this you can "dilate binaries", which add always one pixel to the picture. With "erode binaries" it will be canceled for one pixel. Even with this function not every boundary got visible, so the boundaries have now to be drawn by hand what is shown in figure 2.

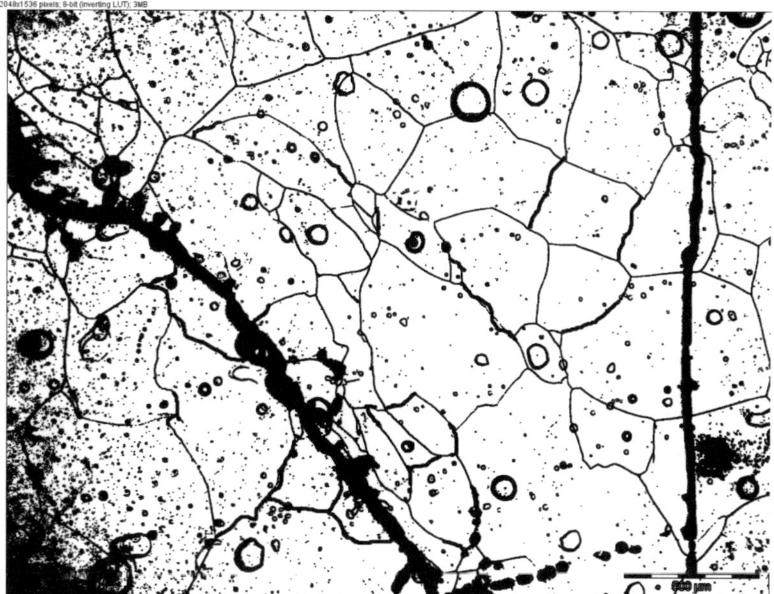

Fig. 2: Picture 430.9-1 after drawing the crystal boundaries with hand.

If the picture doesn't look in an acceptable way, the scale-bar can be removed otherwise it will be measured as a boundary and it can now inverted. How this looks shows figure 3.

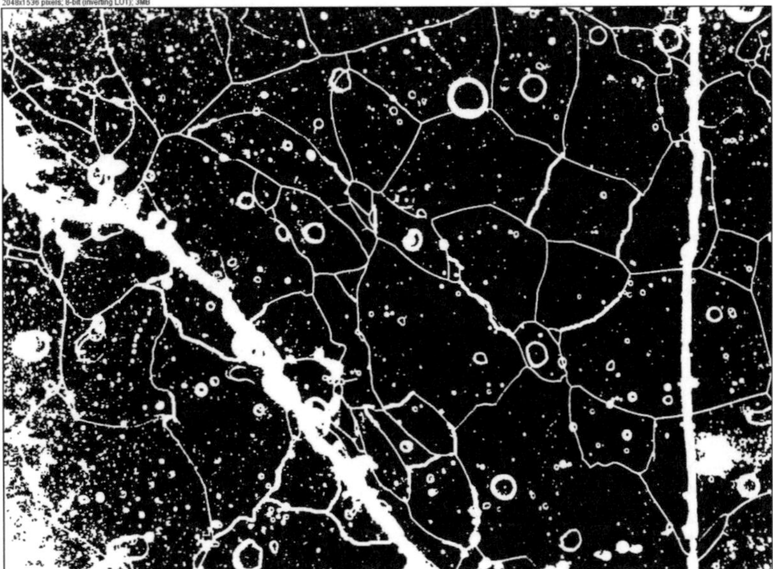

Fig. 3: Inverted picture 430.9-1 after editing.

After the inversion the analysis of the particles can be done. The results are shown in figure 4 and table 1. A potential error source can be seen in figure 4: Gas bubbles under the section and cracks within the sample are also measured as crystal boundaries. To avoid a minimum of errors, the smallest particles, which are mostly bubbles, have to be outlined in the calculation.

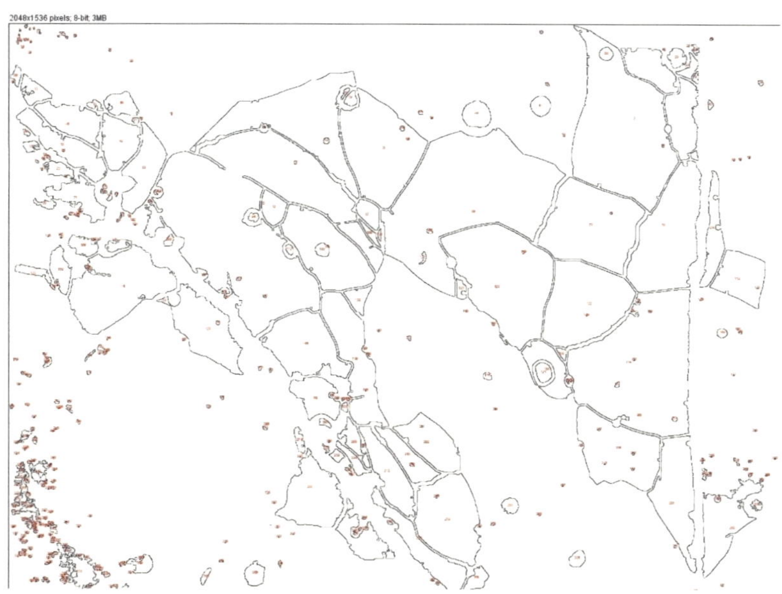

Fig. 4: Particle analysis of picture 430.9-1.

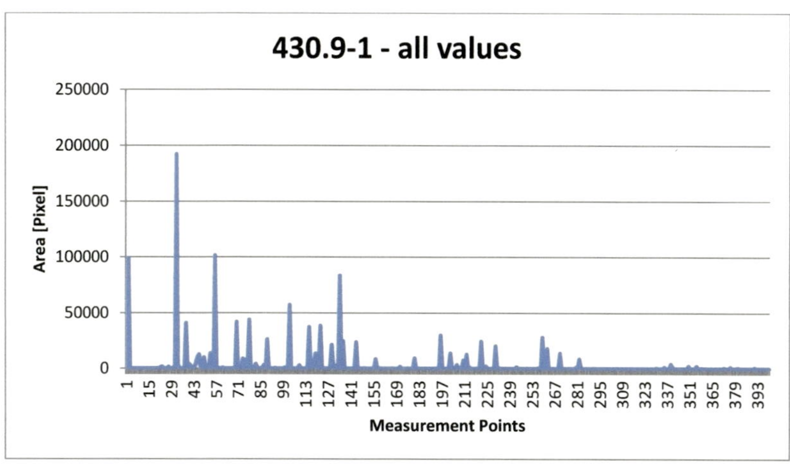

Tab. 1: Measurement results plotted in a diagram. All values including even the smallest points.

The average area is now 3008.7 Pixel. The calculated diameter for the salt pores (assuming that they are round) is 158.76 Pixel. The exemplary calculation can be seen below.

$$\sqrt{\dfrac{\dfrac{average\ area}{0.75}}{\pi}} \cdot 2$$

As you can see in table 1, the inclusion of all measurement points – even the smallest bubbles – causes a big error value, so all pixels smaller than 1000 Pixel were taken out. The new measurement plot can be viewed in table 2.

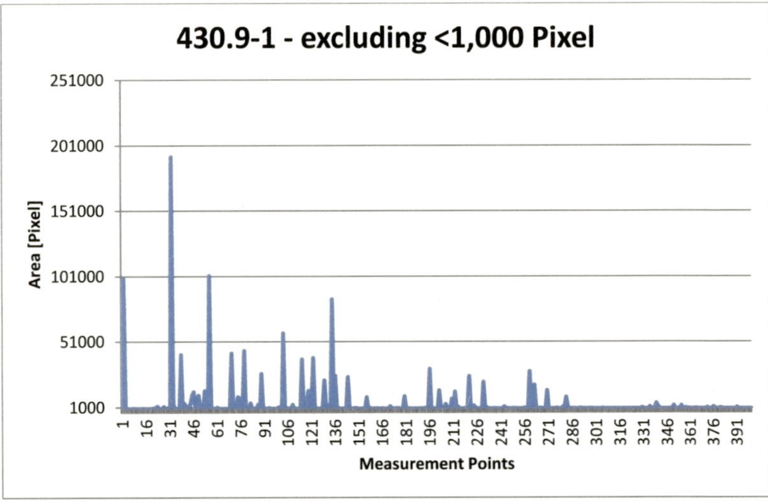

Tab. 2: Measurement results for picture 430.9-1 excluding all grains smaller than 1000 Pixel.

The average area changes now to 17314.44 Pixel and the average diameter of the salt pores to 171.45 Pixel. With this method all other pictures were calculated.

II. Calculation Results of all Pictures

The calculation results summarized for all pictures can be seen in table 3. The plotted diagrams are shown in table 4a-4j.

	Average Area [Pixel]	Average diameter [Pixel]
423.3-2	23385.39	199.25
430.9-1	17314.44	171.45
430.9-2	448741.29	8728.15
430.9-3	15729.16	160.41
443.5-1	26802.43	213.31
443.5-3	19541.19	182.14
443.5-5	48589.57	287.21
443.5-6	31841.26	232.50
457.5-1	15544.49	162.45
457.5-2	21757.06	192.19
457.5-4	11321.51	138.63

Tab. 3: Calculation results for all pictures based on the example mentioned above.

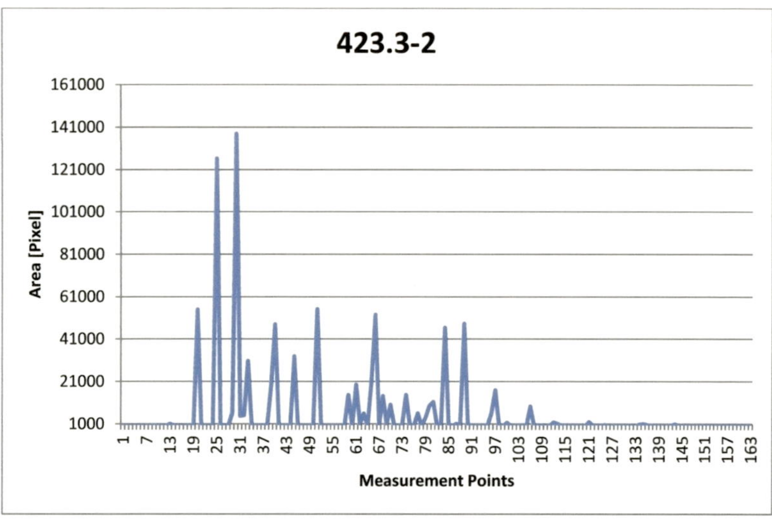

Tab. 4a: Sub grain size measurement for picture 423.3-2 excluding all particles smaller 1000 Pixel.

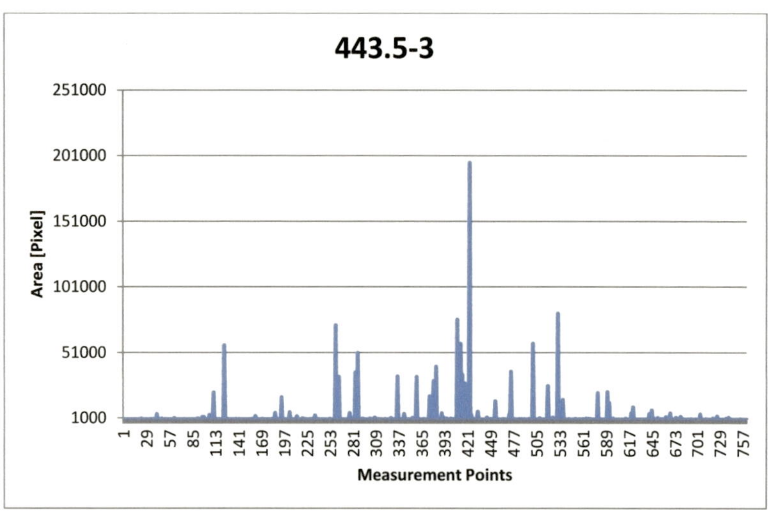

Tab. 4b: Sub grain size measurement for picture 443.5-3 excluding all particles smaller 1000 Pixel.

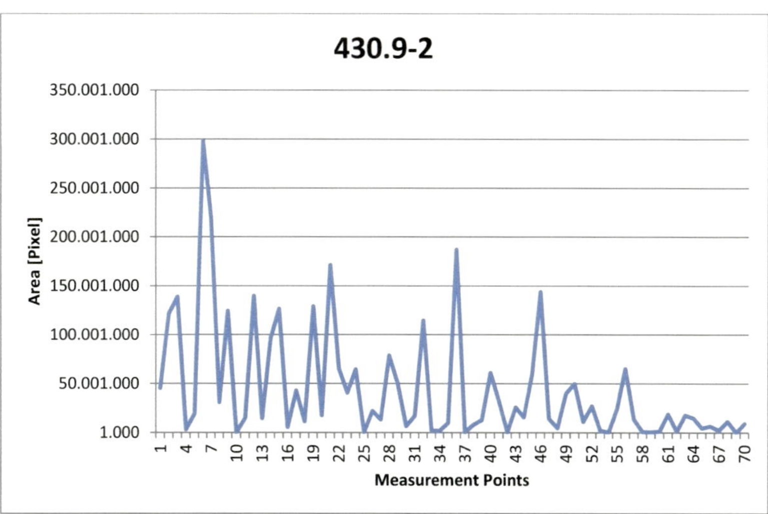

Tab. 4c: Sub grain size measurement for picture 430.9-2 excluding all particles smaller 1000 Pixel.

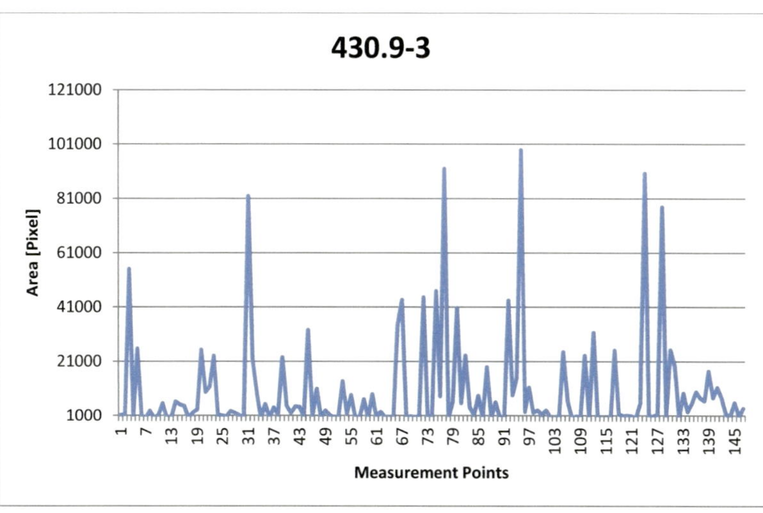

Tab. 4d: Sub grain size measurement for picture 430.9-3 excluding all particles smaller 1000 Pixel.

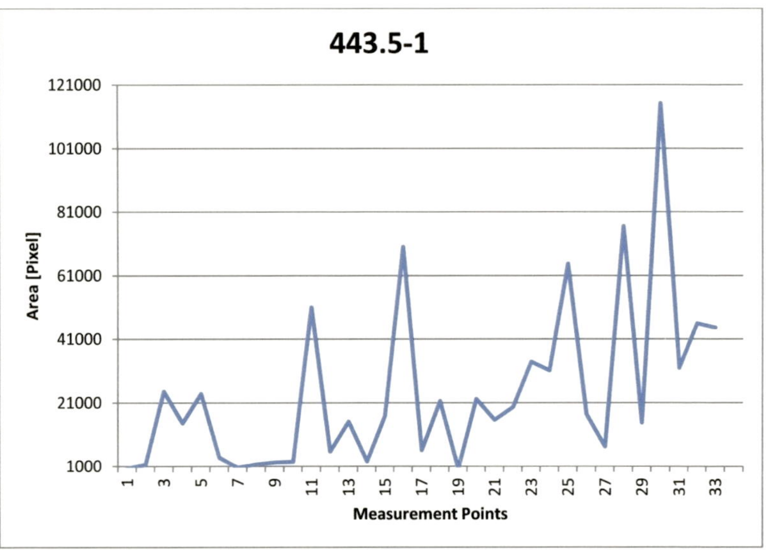

Tab. 4e: Sub grain size measurement for picture 443.5-1 excluding all particles smaller 1000 Pixel.

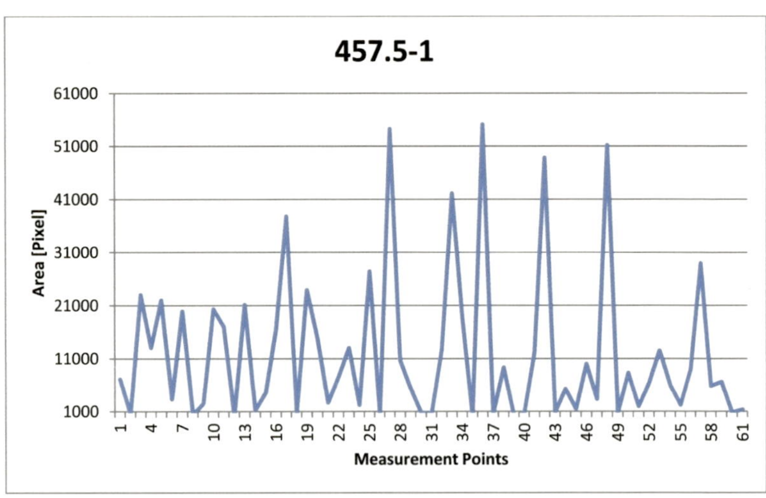

Tab. 4f: Sub grain size measurement for picture 457.5-1 excluding all particles smaller 1000 Pixel.

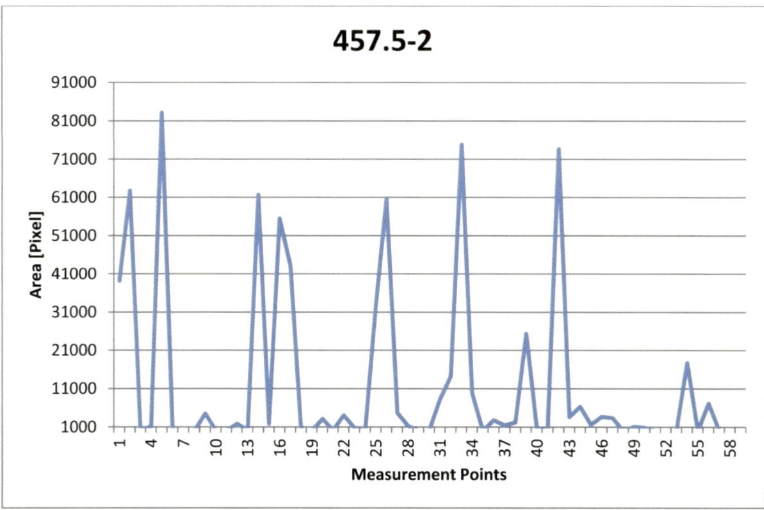

Tab. 4g: Sub grain size measurement for picture 457.5-2excluding all particles smaller 1000 Pixel.

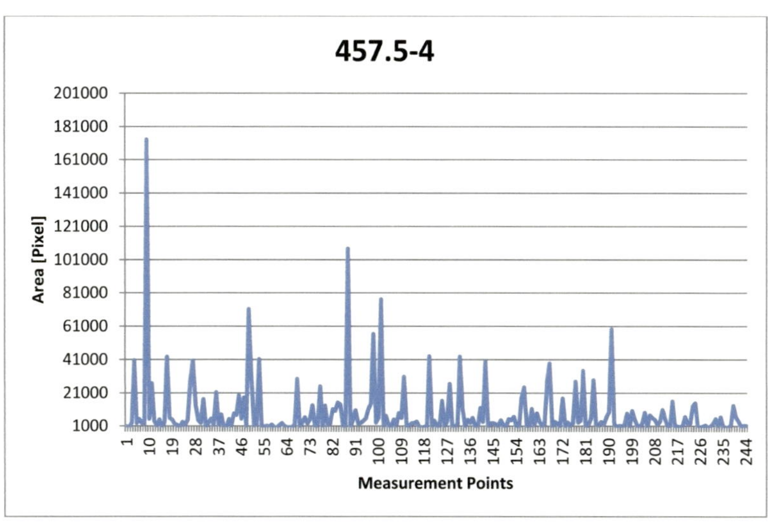

Tab. 4h: Sub grain size measurement for picture 457.5-4 excluding all particles smaller 1000 Pixel.

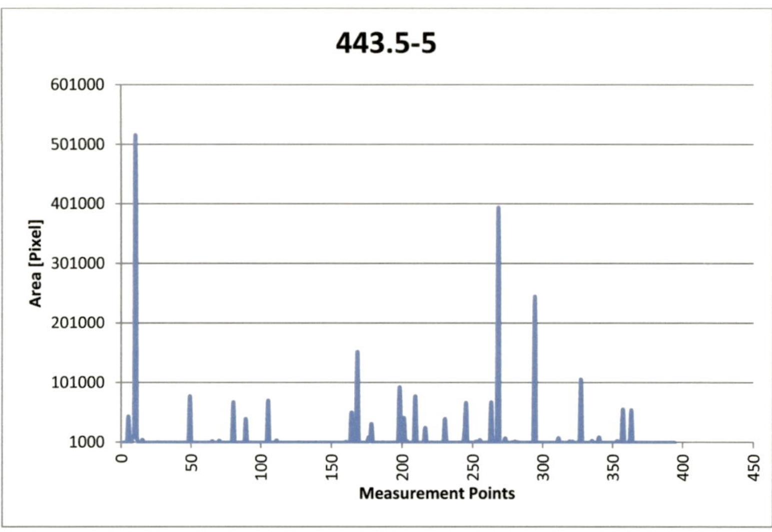

Tab. 4i: Sub grain size measurement for picture 443.5-5 excluding all particles smaller 1000 Pixel.

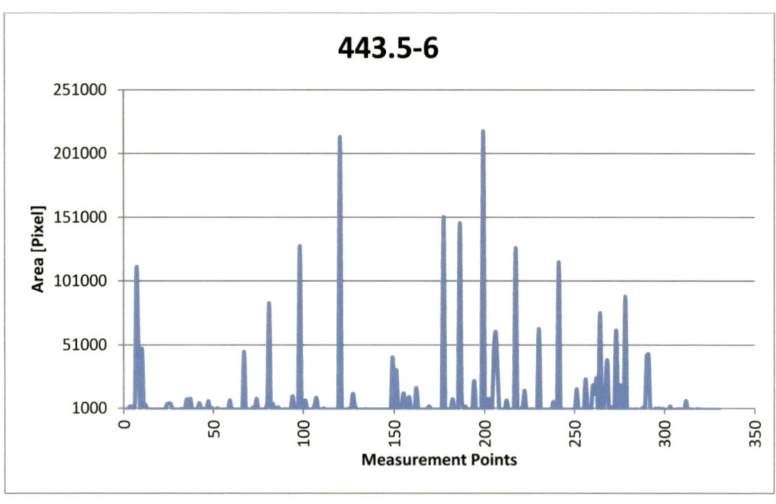

Tab. 4j: Sub grain size measurement for picture 443.5-6 excluding all particles smaller 1000 Pixel.

III. Error Calculation

The error propagation was done with the standard deviation. The results for each picture are shown in table 5.

	Standard Deviation for Average Area
423.3-2	17719.86913
430.9-1	30550.79625
430.9-2	60423905.98
430.9-3	21391.2847
443.5-1	26691.33453
443.5-3	30339.33812
443.5-5	95276.88837
443.5-6	47388.30372
457.5-1	14768.88281
457.5-2	26694.93163
457.5-4	19359.94493

Tab. 5: Standard Deviation calculated for each picture on the average area excluding all particles smaller 1000 Pixel.

The results show that the standard deviation is relatively high for each picture. The cause of this is maybe the mentioned error sources; like gas bubbles (under the thin section) and cracks within the sample are also measured as crystal boundaries.

IV. Palaeo-stress

To compare the sub grain size (average area) with palaeo-stress results by Carter et al. (1982) and Burke et al. (1981) for rock salt [1], first the average area has to be converted from Pixel into µm. Each of the pictures has after inverting and measuring an average size of about 3 MB (= 3,000,000 Bytes) for the .tif files. One of the first steps was a conversion into 8-bit, therefore 1 Pixel has a size of 8 Byte:

$$\frac{300000 \; Byte}{8 \; Byte} = 375000 \; Byte$$

This value is now taken to calculate the grain size [µm] out of the average grain size [Pixel]. A calculation example is done on picture 430.9-1:

$$\frac{375000 \; Byte}{17314.44 \; Pixel} = 21.66 \; \mu m$$

The results for all pictures are shown in table 6. The estimated palaeo-stresses with the given literature [1] is also presented in table 6.

	Sub grain size [µm]	Estimated palaeo-stress [MPa] [1]
423.3-2	16.04	5.7 (after Burke et al. 1981) 11 (after Carter et al. 1982)
430.9-1	21.66	4.5 (after Burke et al. 1981) 9.5 (after Carter et al. 1982)
430.9-2	0.84	Not given
430.9-3	23.84	4.2 (after Burke et al. 1981) 8.9 (after Carter et al. 1982)
443.5-1	13.99	8.9 (after Burke et al. 1981) 7 (after Carter et al. 1982)
443.5-3	19.19	5 (after Burke et al. 1981) 9.4 (after Carter et al. 1982)
443.5-5	7.72	Not given
443.5-6	11.78	4.3 (after Burke et al. 1981) 8.3 (after Carter et al. 1982)
457.5-1	24.12	3.65 (after Burke et al. 1981) 8.9 (after Carter et al. 1982)
457.5-2	17.24	3.75 (after Burke et al. 1981) 10.5 (after Carter et al. 1982)
457.5-4	33.12	2.6 (after Burke et al. 1981) 5.35 (after Carter et al. 1982)

Tab. 6: Average grain size [µm] for each picture.

The higher the sub grain size the lower is the palaeo-stress. For grain sizes below 10 µm the literature doesn't give any palaeo-stresses, but with this known relation it has to be above 10 MPa (after Burke et al. 1981) or above 20 MPa (after Carter et al. 1982). The average palaeo-stress for this rock sample is 4.7 MPa (after Burke et al. 1981) and 8.76 MPa (after Carter et al. 1982).

For all pictures the palaeo-stress varies a lot. This can be caused by the error sources mentioned in the parts above. So this method doesn't allow an accurate determination of the sub grain size and therefore the palaeo-stress of a rock salt.

V. References

- http://www.mineralienatlas.de/lexikon/index.php/MineralData?mineral=Halit (23.01.13)
- [1]

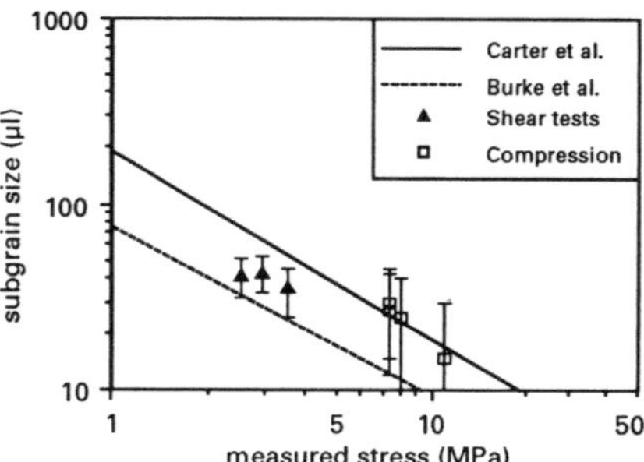

Figure 4.8:
Subgrain size versus applied ($\sigma_{applied}$, $\tau_{applied}$) stress for synthetic rocksalt experimentally deformed in compression (squares) and shear (triangles). The solid lines represent the empirical relationships obtained by Carter et al. (1982) and by Burke et al. (1981).